Glass Steagall versus Economic Collapse

by Rolf A. F. Witzsche

Contents

3

4

About the Illustrated Science series
On the Ice Age and Climate Change
and the book

Glass Steagall versus Economic Collapse
Book 3 of the series, Big Bang Blow Out

When society measures with the meter of entropy, it measures not its potential in boundless development, but measures instead its perceived smallness. On this path it has become too expensive to live, much less to build the 6000 completely new cities for a million people each that are needed to relocate the nations living outside the tropics, into the tropics, in order that humanity may continue to live when the phase shift to the next Ice Age happens, which will likely occur in the 2050s.

The 1933 Glass Steagall law in the USA, was a brave attempt to create an anti-entropic financial system. It had an anti-entropic effect and enabled the USA to develop itself within a couple of decades into the richest nation and the strongest economic force in the world. But this too, needs to be superseded when the building of 6000 new cities and the relocation of much of the worlds agriculture is to be accomplished within the 30 years that we may have still remaining before the phase shift to glaciation climates begins. For this we need an economic system that mirrors the creativity of the universe itself, where every atom is a dynamic construct that is 100,000 larger in size that sum of its parts. That's the nature of anti-entropic economics. If that is what we aim for, the future existence of humanity is assured. If not, the Ice Age Challenge will not be addressed then, and not be responded to, so that the coming phase shift to glaciation conditions in the 2050s will overwhelm humanity. Very few people will then survive the consequences. That's why 3 books are needed to explore the Big-Bang Entropy issues.

Mainstream cosmology regards the universe, the galaxies, and the solar system as exclusively organized by gravitational force that is known to be the weakest universal force. Mass and gravity are all that the Big Bang Theory allows. However, the next higher-order force in the universe is the electric force that is 39 orders of magnitude stronger than the gravitational force. It is expressed in plasma that makes up 99.9999% of the universe. This reality is not allowed to be recognized as an organizing force in the universe, because it is expressed in electrically charged plasma that is deemed not to exist. Cosmology thereby imprisons itself with cosmo-mythologies where nothing is actually true, and humanity becomes imprisoned with it, by the false concepts. One of biggest imprisoning fudge factors, is the Big Bang theory itself.

While technology has furnished astronomy with amazing capacities for looking at the universe, ironically, what is observed is being falsely interpreted on the basis of assumptions that are simply not true, that are mythological assumptions. As a consequence, ironically, mainstream astronomy looks at the universe blindfolded. What comes out of it, of course, are tragic misperceptions. The results are often so confusing that mysterious fudge factors need to be invented to make the results appear plausible. No such fudge factors are needed in plasma cosmology.

With the next Ice Age on the near horizon, potentially beginning in the 2050s, we cannot afford to play games with fudge factors. The recognition of the true nature of the universe, the galactic system, and the solar system, that together drives the Ice Age dynamics, becomes an existentially critical issue. If humanity remains 'asleep' on this front, we may all die in the easy chair of the consequence when the glaciation conditions resume, which evidence promises, will happen quickly.

Plasma in the physical universe is as challenging in perception as the spiritual domain in the human sphere. Both are invisible, except by their effects, but they are understandable and knowable. But how does one break away from the fairy tales that inspire delusions? Answers must be found.

With the Ice Age Challenge now before us, we face two imperatives. One is to understand the real physical dynamics that power and affect the Sun, and with it to create the physical infrastructures that enable human living to continue in an Ice Age climate. The second challenge, and this is the greater challenge, is to raise up our humanity to such height as will impel us to get the job done. Some say that miracles are needed on both fronts. But what of it? Are we, as human beings, not the miracle makers on the Earth?

In the real universe, the cosmic operations are anti-entropic in nature, and expanding and progressing. We, ourselves are evidence of this progression. Should this progression have ended? Neither is our Sun isolated from the progressive nature of the universe, but expresses its dynamics, its resonating plasma streams, and their reflection in the climate on Earth. Shouldn't we develop ourselves spiritually and culturally, likewise?

Climate Change reflects the nature of the universe. It should also be reflected in us.

The Earth itself is the creation of the Sun, with its atoms having been massively synthesized in high-energy times near the center of the galaxy.

The synthesizing plasma fusion is presently at a low state, though it is currently enhanced for our Sun by electromagnetic 'Primer Fields' that focus interstellar plasma onto the Sun in a highly condensed manner. When the plasma-focusing system becomes inactive, below the required threshold conditions, the Sun reverts to a type of cosmic default level with 70% less energy being radiated, and higher rates of solar cosmic-ray flux being experienced.

At the present rate of plasma diminishment being experienced, the solar activity phase-shift threshold to the next Ice Age period may be crossed in 30 years, or in the 2050s, most likely. With the primer-fields system gone inactive by then, the climate on Earth will get 40 times colder than the Little Ice Age in the 1600s had been. Ice core evidence promises that. Without the needed preparations for human living in such an environment, 99% of humanity would die of starvation, both by the cold, and by CO_2 depletion that diminishes agriculture, as more CO_2 becomes dissolved into the sea.

With the 'Primer Fields' being critical for our very existence, the exploration of them is likewise critical.

In the Little Ice Age, between 10% and up to 30% of the populations in Europe had perished by starvation. The last Big Ice Age was evidently vastly harsher. Only 1-10 million people emerged from it alive. That's all we had after 2 million years of development. We want to do far better this time around; and we can, with large-scale technological infrastructures for our food supply. But will we create them? Will we get the job done in the 30 years that we still have left before the Ice Age starts anew? Will we even consider it? And how certain are we that the phase shift to the next glaciation period will begin, as the evidence suggests, in the 2050s? We have no slack on this front. Should we fail us on this absolute front, we would be committing suicide.

Numerous fields of evidence tell us that the next Ice Age is near. That's where the truth begins. Most of the evidence was discovered in the 1990s and thereafter. Some evidence is measured in ice cores; some is measured in space, by satellites. Some measurements are also made on the ground in terms of measurements of the Earth's magnetic-pole drift observed in northern Canada. All of this is seen combined with high-energy physics experiments at a leading national laboratory, and is also explored in the small in static experiments.

So, what will the answer be? Will we move with the evidence? Or will we lay ourselves down to die by default?

It takes an independent researcher to brake the taboos that have kept mainstream cosmology imprisoned, increasingly, during the past century, even while what is regarded as taboo is known to be wrong.

The Illustrated Science series is intended to open the scene beyond the threshold of accepted taboos, to where the actual physical evidence speaks for itself.

The scope of the existential challenge that the Ice Age brings with it, takes astrophysics out of the academic domain and places it into the foreground as one of the most-critical issues of our time. The big Climate Change events that have already worldwide effects are mere fringe effects in the flow of the ever-changing cosmic dynamics. The big effect, when the Ice Age begins anew, promises to be caused by a dimmer and colder Sun. The loss of 70% of the Sun's radiated energy defines our climate future that begins in the near term.

Sure, we can live with all that by creating new platforms for agriculture that are able to operate under Ice Age conditions. But will we do it? The task is enormous. Or will we fail ourselves on this front? We have no reason to allow us to fail. We have the materials and energy resources on hand to accomplish everything that is required for us to continue to live in an Ice Age World. But will we do it? The big question that never goes away, therefore, is; will we develop our inner resources as human beings sufficiently to get the job done, and to get it done in time? Or will we do nothing, ignore the challenge, and condemn our children and one-another to an agonizing death by starvation? That's the choice.

Towards meeting the inner challenge, I have created the epic series of novels, The Lodging for the Rose. And further, towards meeting the science challenge, I have produced numerous research books and several dozen exploration videos that the Illustrated Science series is modeled after. The work is the result of a quarter century of research, for which numerous elements of evidence in related fields came to light during the timeframe of my research.

It is my hope that the work that went into all of these projects will help in some degree - for humanity that we are all a part of - to write itself a ticket to have a future.

High-resolution color images, of the images in this book, can be obtained at www.iceagetheatre.ca

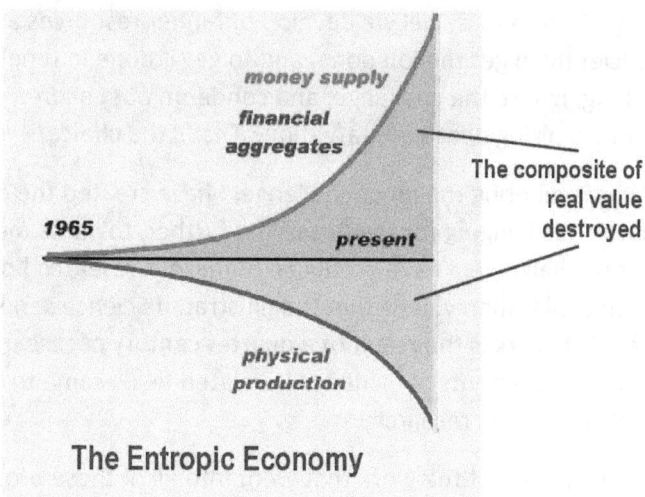

The Entropic Economy

The colonial rule for stealing the wealth of nations, does actually precede 'Adam Smith', who may have tried to hide it. It appears to have been recognized quite early in the history of civilization that the process of stealing, in whatever form it may have been carried out, collapses the thereby ravished economy, so that the stolen wealth itself becomes worthless thereby for the simple fact that monetary values stand as a claim against a productive economy, which under the regime of looting becomes a destroyed economy.

The Colonial Age,

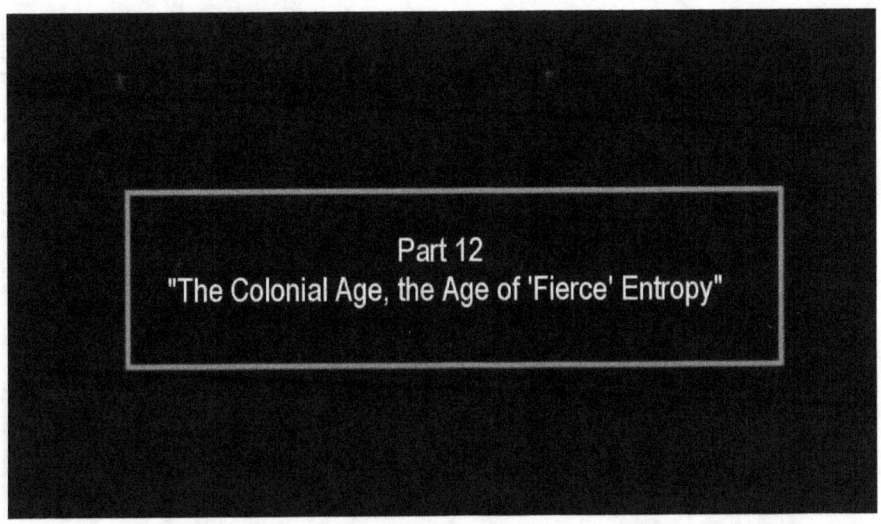

"The Colonial Age, the Age of 'Fierce' Entropy"

To circumvent the inherent entropic collapse

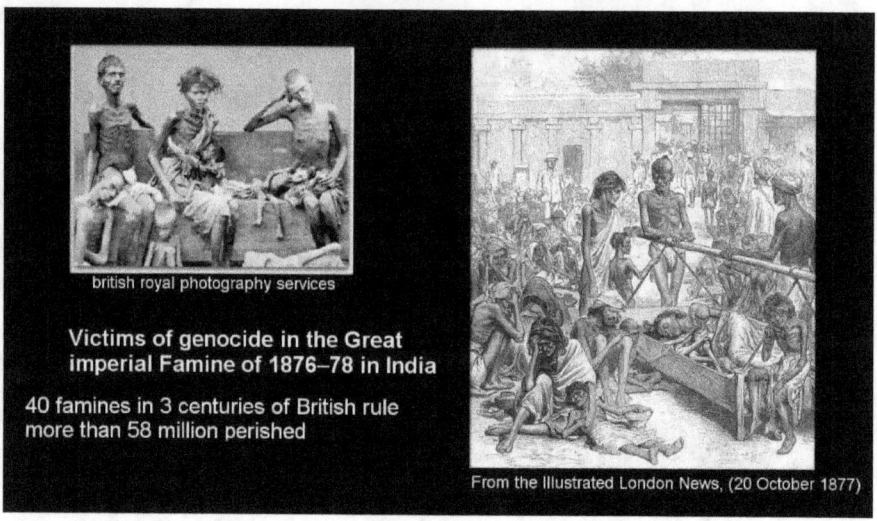

british royal photography services

Victims of genocide in the Great imperial Famine of 1876–78 in India

40 famines in 3 centuries of British rule more than 58 million perished

From the Illustrated London News, (20 October 1877)

It may have been in the attempt to circumvent the inherent entropic collapse that is an unavoidable feature of the thievery system, that the masters of the greedy began to spread their thievery across the world to other nations, to ravish them instead of their own local society that the masters found themselves obliged to maintain as a necessary infrastructure for its wars.

The colonial age began

With their success in spreading their looting 'enterprises,' the colonial age began, which of course was enforced with the infamous gunboat 'diplomacy,' and if there was resistance, with war to enforce the raping.

When the resistance was internal

El Tres de Mayo, by Francisco de Goya - Wikipedia

And when the resistance was internal, genocide was unleashed to destroy the hart of of whatever society had dared to raise it head in defiance.

Almost the entire world became subjugated

Colonial Empires of the World 1914

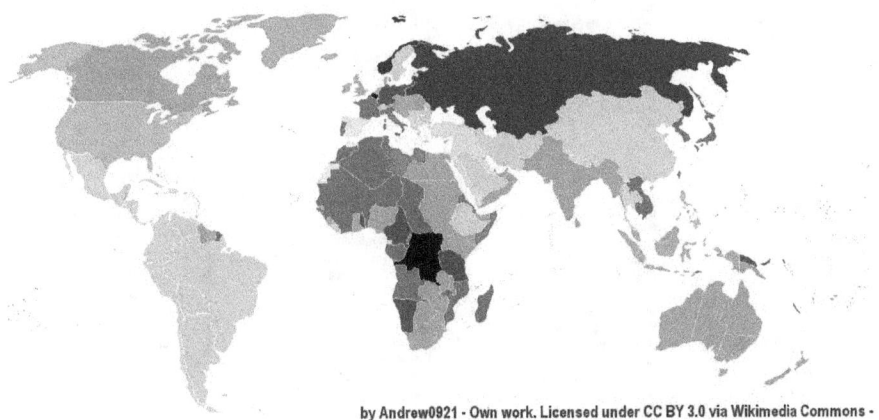

by Andrew0921 - Own work. Licensed under CC BY 3.0 via Wikimedia Commons -

During the colonial age, almost the entire world became subjugated in various ways to tiny groups of oligarchic money bags and royal houses in the 'empire lands,' the lands of the private masters who have looted the world far and wide for their pleasure, by whom humanity was treated as a type of animal to be 'harvested' in the extreme, and discarded at will.

Lists of the subjugated nations

———— British colonies ————

Aden	Gilbert and Ellice Islands
Anglo-Egyptian Sudan	Gibraltar
Ascension Island	Gold Coast
Australia	India
Australian Antarctic Territory	(including Pakistan and Bangladesh)
Christmas Island	Ireland
Cocos Islands	Jamaica
Norfolk Island	Kenya
Bahamas	Malta
Basutoland	Newfoundland
Bechuanaland	New Zealand
British Antarctic Territory	Cook Islands
British East Africa	Niue
British Guiana	Ross Dependency
British Honduras	Tokelau
British Hong Kong	Nigeria
British Malaya	North Borneo
British Somaliland	Northern Rhodesia
Brunei	Oman
Burma	Papua
Canada	Sarawak
Ceylon	Sierra Leone
Cyprus	Southern Rhodesia
(including Akrotiri and Dhekelia)	St. Helena
Egypt	Swaziland
Falkland Islands	Trinidad and Tobago
Fiji Islands	Uganda
Gambia	South Africa

———— French colonies ————

Algeria	Upper Volta	
Clipperton Island	Guadeloupe	
Comoros Islands (including Mayotte)	Saint Barthélemy	
French Guiana	Saint Martin	
French Equatorial Africa	La Réunion	
Chad	Madagascar	
Oubangui-Chari	Martinique	
French Congo	French sMorocco	
Gabon	New Caledonia	
French India	Saint-Pierre-et-Miquelon	
French Indochina	Shanghai French Concession	
Annam	Tunisia	
Cambodia	Vanuatu	
Cochinchina	Wallis-et-Futuna	
Laos	Not shown here:	
Tonkin	Russian colonies	
French Polynesia	German colonies	
French Somaliland	Italian colonies	
French Southern and Antarctic Lands	Dutch colonies	
French West Africa	Portuguese colonies	
Benin	Austro-Hungarian colonies	
Côte d'Ivoire	Belgian colonies	
Dahomey	U.S. colonial possessions	
Guinea	Chinese colonies	
French Sudan	Ottoman colonies	
Mauritania	Japanese colonies	
Niger		
Senegal		

The lists of the subjugated nations, countries, and areas, were long. They contained many names of countries and people and areas that became 'property.'

On this page, only the names of the two biggest groups of colonial properties are listed. These are the names of the British colonies, and of the French colonies. The remaining eleven colonial owners are listed without their properties displayed.

Countless wars were fought throughout the centuries to subjugate the colonies, and to keep them in line.

The American republic was born

The Second Continental Congress voting on the United States Declaration of Independence

July 4, 1776

The American republic was born when a group of colonies took a stand against being subjugated by the global thievery system. The American patriots claimed the inherent freedom of the human being for themselves. They claimed their freedom on the recognition of the inherent anti-entropy in the human society that came to light through its scientific and technological development on the distant shores of America, far from the choking effect of Empire. The patriots discovered that a society's inner development invariably increases its creative and productive power. The discovered anti-entropy of their humanity gave the patriots the dignity with which they stood tall, and said to empire: No More, Never Again.

America did attain its freedom, and fought to retain it

Washington Crossing the Delaware, in 1777 -
"These are the times that try men's souls," (Thomas Paine)

On this basis America did attain its freedom, and fought determined to retain it against all the forces of empire that promptly waged a war to reverse the unfolding history of humanity.
America stood its ground for 130 years.

America lost itself again to the devil within

America lost itself again to the devil within, to the theory of entropy, to the platform for stealing that invited the Federal Reserve Act, the platform for war. It never freed itself from this defeat, but was drawn into world wars by the process.

Colonial wars soon became world wars

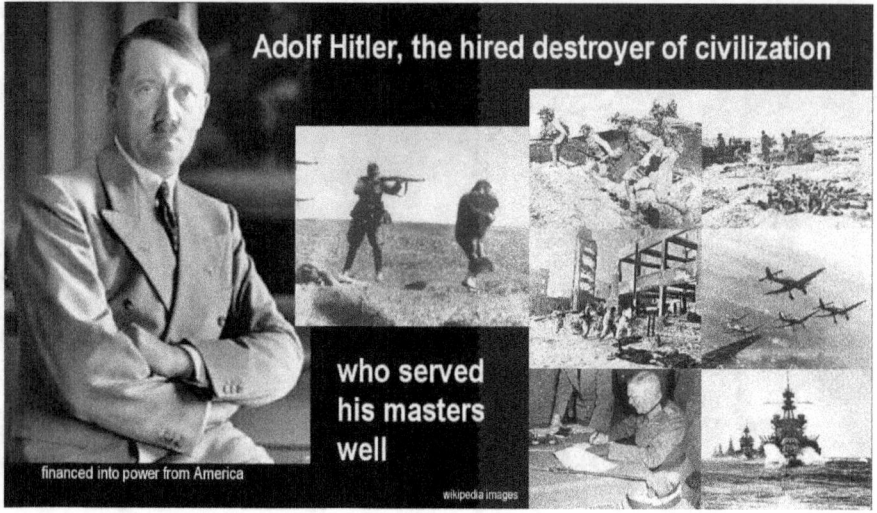

Adolf Hitler, the hired destroyer of civilization

who served his masters well

financed into power from America

wikipedia images

The historic colonial wars soon became world wars, poised to become nuclear war. The buzz words, "First Strike," "Limited," "Pre-emptive," are now applied to nuclear war against the whole world, just as Hitler had applied these terms to his grand schemes of madness in obedience to his masters who had set him up as their puppy dog.

Stealing with the force of 'invincible' arms

We wield weapons because of impotence

While the face of war has been radically modernized in recent years, the platform for war remains the same nevertheless. The masters who stand behind all the platforms for stealing, cling tenaciously to their mistaken belief in universal Entropy for which their devil in the mind inspires stealing with the force of 'invincible' arms, while the masters themselves never go to war to die in the conflicts they create. Soldiers are used for that.

Few soldiers in history knew

UH-1D helicopters airlift members of a U.S. infantry regiment, 1966 - James K. F. Dung, SFC, Photographer

Few soldiers in history knew, or even wished to know, that they laid down their life to die in the dust of some foreign lands for the profits of the wealthy who own them on their strings, and who thereby render them to become mass-murderers before God in the wars that demand them to discard their humanity as a worthless impediment.

War has become a worldwide disease

Russian made HIND Mi-24 helicopter

War has become a worldwide disease that is waged for the purpose of looting, for which enormous resources are wasted in the defence against it.
Sadly, in the heat of the raging battles, the root for the disease, the belief in entropy, becomes largely forgotten. Only the blood remains real that flows into the sand.

Stealing by all means possible, humanity is doomed

Kepler Today - Part 2:

Sovereignty versus Nuclear War

509th Composite Group - B29 aircraft immediately before the bombing mission of Hiroshima Photo by Harold Agnew

For as long as the historic, entropic game of empire continues, which demands stealing by all means possible, humanity is doomed to its self-destruction by war. Without science raising up the sovereignty of our humanity to higher levels, above the small-minded historic game, our hope is slim.

As stealing demands evermore wars

Hiroshima after the bombing

wikipedia

For as long as stealing demands evermore wars, humanity denies itself to have a human future.

The whole of humanity is now doomed

Annihilation is assured

500,000 times
Hiroshima
in one hour

Castle Bravo - the first U.S. test of a dry fuel thermonuclear hydrogen bomb – March 1, 1954 at Bikini Atoll, Marshall Islands

The whole of humanity is now doomed by its belief in entropy that invites stealing. The result is exceedingly tragic. The human world cannot survive the force of half a million Hiroshimas.

Hiroshima
victim of a
terror campaign

wikipedia
Picture courtesy of the National Archives and Records Administration

With the formulation of the cosmic Big Bang theory, the masters of the system of empire attempted to give their deadly and destructive system of entropy, a noble face to justify their terror. With the Big Bang theory, the masters attempted, not to heal their inhumanity, but to give the increasingly horrific destruction of civilization, that their greed demands, a scientific excuse.

Science complied

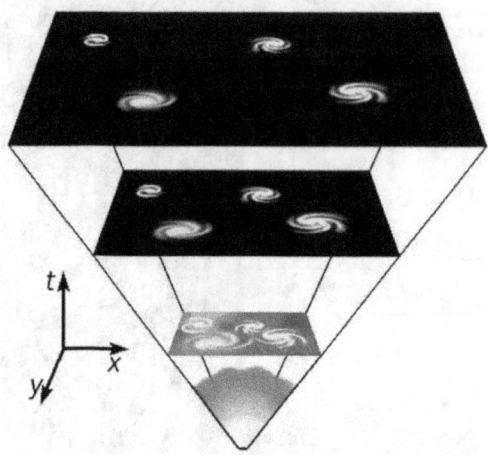

Science complied. It pinned the ugly face of empire, the face of entropy that trails out into nothing, unto the face of the universe, even unto the face of God.

That's an old trick that the empire of Rome had used, and all empires thereafter under the rubric of the "divine right of kings." While the result, that became the cosmic Big Bang theory, is evidently false, as no real evidence supports the theory of the declining and self-consuming universe, the theory's myth of universal entropy nevertheless still rules society powerfully. It rules it as a philosophy that upholds the dream of entropy, a dream where all energy is consumed into nothing, which is the key feature that the world empire cannot avoid, with it being built on the platform of entropy.

Thievery inherent in the kingdom of empire

The disease of thievery as a system that is inherent in the kingdom of empire, has unfortunately become so powerful by it's being promoted as a science, so that the budding opposite political structures that are based on the platform of anti-entropy, the platform of human self-development and self-protection, have become largely eliminated from the landscape of civilization in the modern, neo-colonial, imperial world, for as far as its tentacles have been able to reach.

A new wind is rising in the distant lands

Fortunately the disease is fading. A new wind is rising in the distant lands where the tentacles of empire are loosing their grip. The winds of healing are rich with scientific development, cultural optimism, infrastructure building, industrial production, energy development, cultural development, which are all features of human development. These winds flow from China today, and Russia, and India, who have become the pioneers for a new hope for humanity.

Where America had stood when it stood tall

These pioneering nations, and those who joined hands with them, stand at the forefront today in the race away from empire, where America had stood when it stood tall, before it became re-colonized again.

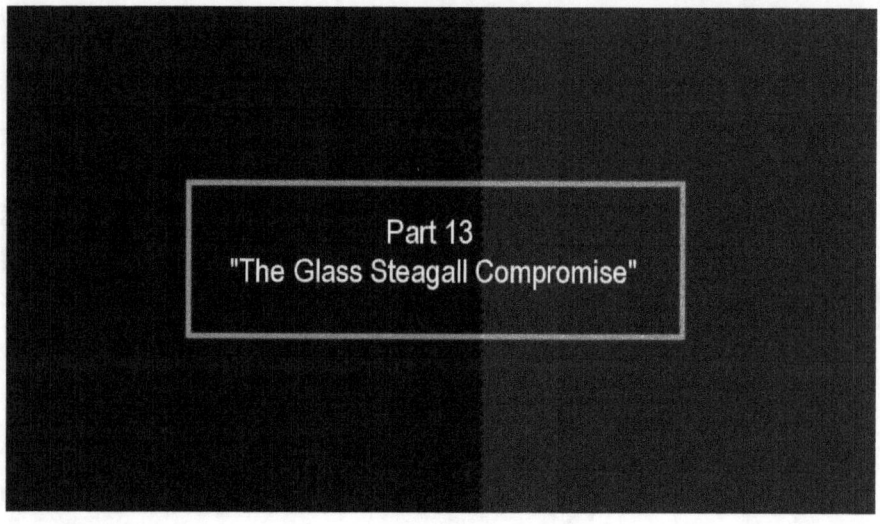

"The Glass Steagall Compromise"

With the repeal of the Glass Steagall Act

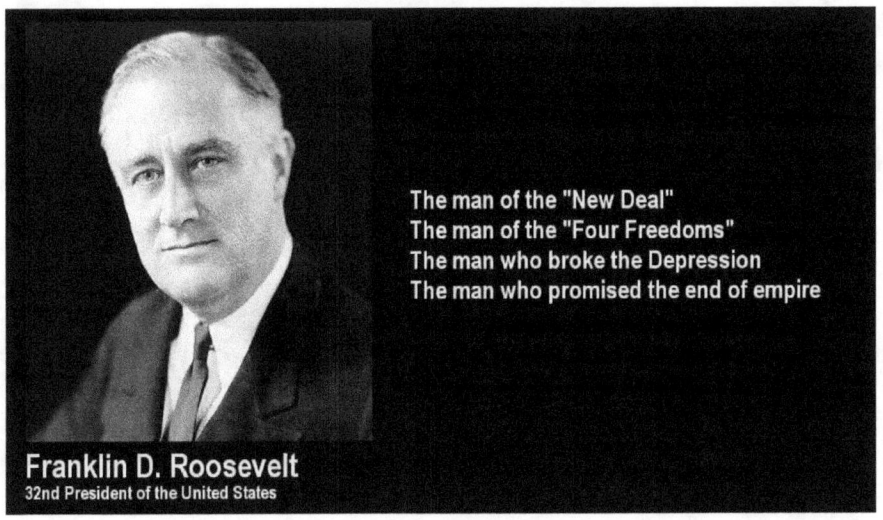

The man of the "New Deal"
The man of the "Four Freedoms"
The man who broke the Depression
The man who promised the end of empire

Franklin D. Roosevelt
32nd President of the United States

In America the destruction of the nation in the neo-colonial world, was achieved with the repeal of the Glass Steagall Act. The Glass Steagall legislation had been set up during the Franklin Roosevelt Presidency. It was one of the President's first steps to enable America to pull itself out of the economic slum of The Depression that had resulted from the entropy of the looting of society.

The new colonialism of the Euro empire

European Central Bank
in Frankfurt - Germany

In Europe, the same type of destruction of historic economic
culture, was achieved with the new colonialism of the Euro
empire. The Euro system effectively cancelled all historic structures
of the European nations' self-protection and self-development.
In both cases, the grand stealing from the respective nations via
various bank-bailout mechanisms, was not only legalized with
legislation, but was, in the case of Europe imposed by cleverly
arranged treaty obligations, such as the Lisbon Treaty that was
designed to be a trap.

The freedom to steal has become protected

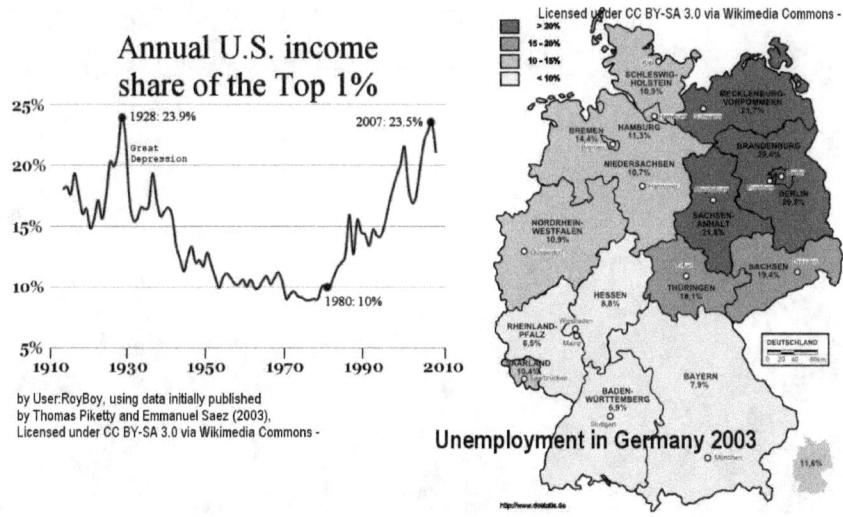

Annual U.S. income share of the Top 1%

1928: 23.9%
2007: 23.5%
Great Depression
1980: 10%

by User:RoyBoy, using data initially published by Thomas Piketty and Emmanuel Saez (2003), Licensed under CC BY-SA 3.0 via Wikimedia Commons -

Licensed under CC BY-SA 3.0 via Wikimedia Commons -

Unemployment in Germany 2003

In both areas of the world, Europe and America, the freedom to steal has thereby become protected, while society's self-development has become prohibited.

Both, America and Europe, have so totally bankrupted themselves in this process, as members of the new Flat Earth Society's kingdom of entropy; that preparations are now being made towards the next world war to capture Russia and China as the final remaining resource in the world for continued stealing.

War against Russia and China

Annihilation is assured

500,000 times
Hiroshima
in one hour

Castle Bravo - the first U.S. test of a dry fuel thermonuclear hydrogen bomb - March 1, 1954 at Bikini Atoll, Marshall Islands

The resulting war against Russia and China, that is now being
prepared, will invariably become a nuclear war. All studies have
shown this. And the studies have shown too, that nuclear war is
unsurvivable.

In America, the Glass Steagall legislation was repealed in 1999

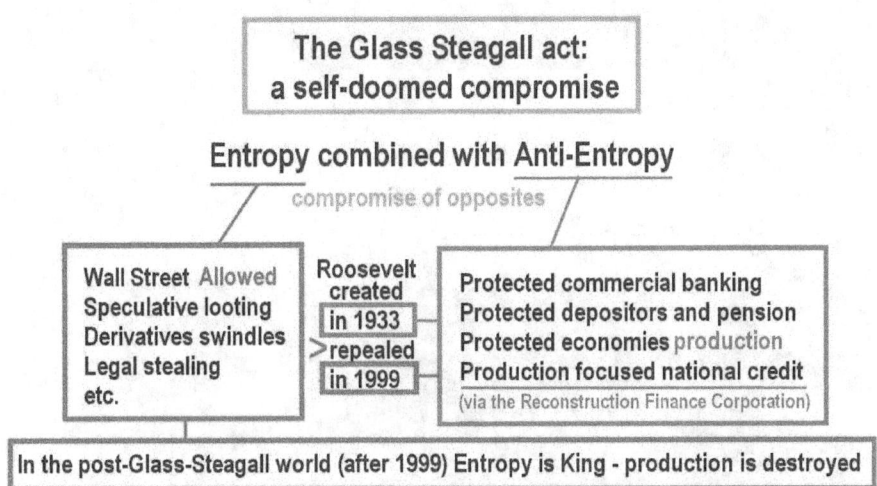

In America, the Glass Steagall legislation had once furnished a type of anti-entropic platform that stood for 67 years and served national development. With it, America had achieved the highest level of general prosperity of any country in the world. It was this foundation for prosperity that was repealed in 1999.

As one might expect, the repeal of the law that had prohibited stealing, opened the flood gates to great national tragedies. Ironically, the resulting tragic failure in civilization was almost assured from the outset. It was assured, because the legislation had been set up as a compromise. The law had compromised, in that it had merely separated the entropy of empire, the kingdom of stealing, from the anti-entropic productive platform that had furnished national self-development and self-protection. It is here, in the fundamental compromise, where its failure is rooted.

Compromise on principle became its doom

1933 - the Glass-Steagall law to protect that value of the U.S. currency - now repealed

The Glass Steagall law had allowed the entropic platform of the kingdom of stealing, to continue in the background as a compromise. The compromise on principle became its doom. One cannot operate a compromise that incorporates two opposite platforms. The kingdom of stealing named Wall Street, and the principle of national development for Main Street, are irreconcilable. Stealing is destructive. It doesn't create and develop anything.

Entropy and Anti-entropy are mutually exclusive

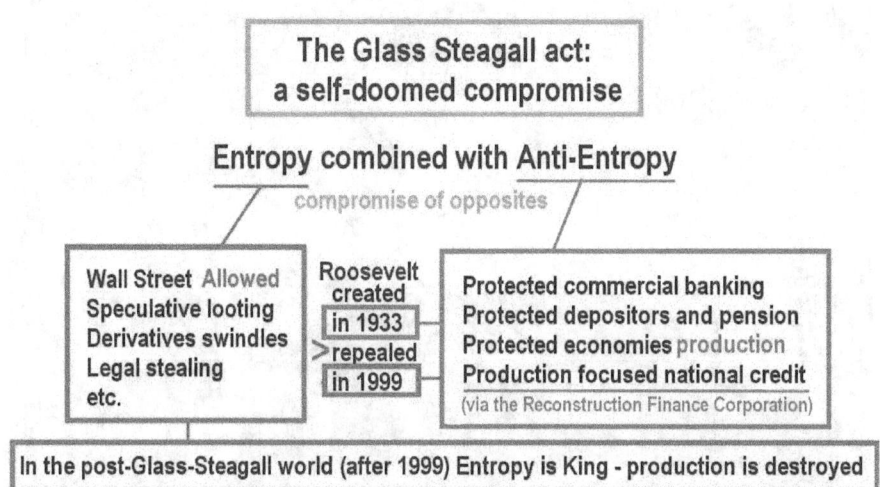

In the post-Glass-Steagall world (after 1999) Entropy is King - production is destroyed

Entropy and Anti-entropy are mutually exclusive platforms. Entropy is not the law of the universe. The universe is Anti-entropic in principle. It is exclusively self-powered and universally self-developing. Diminishment is not a feature of the universe. The universe is forever developing from its infinite, continuing, all-pervading source, and is forever increasing and improving its dimensions in all aspects. Nothing is winding down in the universe. No form of stealing is happening.

The theory of self-consuming stars

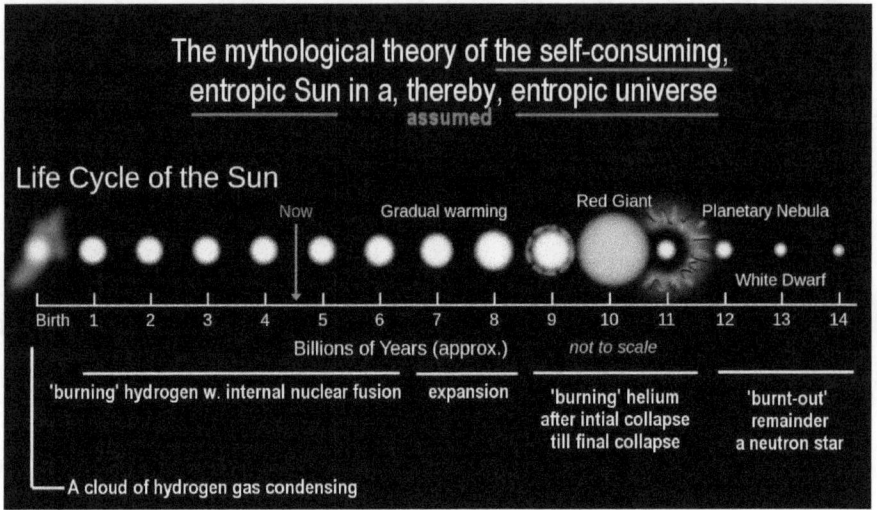

The theory of self-consuming stars that die in the end by their energy depletion, is a myth. The same myth has been falsely applied to humanity and civilization. It is a tragic folly for humanity to reject the anti-entropic platform that the universe operates on, and devise for itself an opposite platform based on a theory that is false in every respect.

The Glass Steagall Act failed because of this cultivated folly of accepting entropy as a quality of natural dynamics. Glass Steagall failed, because the folly of imposing entropy into the premise of civilization had been left unresolved. If the belief in entropy had been overcome, empire would have been closed down, and the development of the USA would have continued.

But by its compromising on a fundamental principle, the Glass Steagall Act has stood effectively in self-denial from its very beginning. This self-denial eventually opened the door to its doom.

We play the same compromising game

We play the same compromising game by allowing nuclear war to stand in the world as an option for warfare. It is an element of the entropic kingdom of stealing, an element that has remained unresolved. This element cannot be resolved by itself in isolation. The ever present nuclear war danger can only be resolved by scrapping the entire package of empire that it is a part of. This means stepping up to the anti-entropic platform for civilization, which is the natural platform for humanity.

The Glass Steagall compromise

The man of the "New Deal"
The man of the "Four Freedoms"
The man who broke the Depression
The man who promised the end of empire

Franklin D. Roosevelt
32nd President of the United States

It appears that the Glass Steagall compromise was not made entirely by choice in 1933. It was allowed by default, because the new President, President Roosevelt, with all the popularity that he had, with which he had secured his election, couldn't muster the political backing in the house of government to eliminate the entropic platform of financial thievery from the national landscape. The compromise came from that. In like manner, the political backing is still far out of sight to banish nuclear war that looms as the final stage of political entropy.

Twenty years before Roosevelt

Washington D.C. building of the U.S. Federal Reserve - a private central bank

By the foolish compromise that was allowed in 1933, the kingdom or entropy had survived and become a monster. The monster should have been cast out right then. Instead it was permitted to continue as the private owner of the financial house of the USA in the form of the Federal Reserve system.

The tragedy had it origin long before Franklin Roosevelt's time. Twenty years before Roosevelt the entropic kingdom of thievery had already corrupted the nation so deeply, that it was able, with corrupted politicians, to steal the nation's currency into private hands, by an operation that became deceptively named a "Federal" system. From this lavish base of complete control over the nation's money, which the Glass Steagall Act had continued, the nation's leaders were subsequently further corrupted to repeal the Glass Seagall Act - the very platform on which the nation's prosperity had been built, in order to get the nation back to the Depression that the Federal Reserve system has created shortly after it was formed.

This immense corruption of the U.S. Congress and Senate, to repeal

Glass Steagall, was achieved in 1999 with a giant slush fund from the kingdom of money amounting to hundreds of millions of dollars. These huge sums, handed out under the table to a selected few, did buy the terminating votes from the small-minded who are easily turned to become traitors for hire.

The year 1999 marks the historic beginning

Since 1999

President Bush elected (the war and terror President)
The September 11, 2001 State terrorism event
'Perpetual' War against Afghanistan (2001 till end of 2014)
Iraq War (2003-2011)
Legalizing of Torture
Collapse of the auto industry
Home foreclosure crisis
War threats against Iran
President Obama elected (the shutdown President)
Great bailout Bank Heist ($50 trillion stolen since 2008)
Healthcare and social security decimation
Libya color revolution (war) - (2011 - to the present)
Egypt color revolution (war) - (2003 - to 2004)
Syria color revolution (war) - (2012 - to the present)
Ukraine violent overthrow as a step towards Russia
Nuclear War threats against Russia and China
with the western financial system dead on its knees.

...we've stood closer to nuclear war
than ever before, and more often.

The Kingdom of Entropy
the post-Glass-Steagall era
1999 to the present
an era of Extreme Stealing
Terror, and Perpetual War.

Annihilation is assured

500,000 times
Hiroshima
in one hour

The rest is history, as people say. This history is still unfolding. The year 1999 marks the historic beginning of what may be called one day, the greatest national tragedy in the entire existence of the USA.

The tragedy started with collapsing financial values that prompted the infamous 911 State terrorist event, which in turn prompted the 'Perpetual' War doctrine, supposedly to fight terrorism. On this track war was brought to Afghanistan, then Iraq, with the legalization of torture along the way. At home in the USA, the collapse of the auto industry decimated employment, with war threats against Iran occurring in the background, while the home foreclosure crisis unleashed social chaos.

Since this didn't solve anything on the financial front, the Greatest Bank Heist in the history of the world was staged seven years later, this time not to rob the banks, but to rob society of upwards to $50 trillion to bail out the gambling casinos that the banks had become. Of course, to save money, society's healthcare and social security system was decimated, which is still ongoing.

Since this immense sacrifice didn't solve the financial collapse crisis either, the scourge of war was brought against Libya to murder its leader in the name of liberty, to liberate its oil resources. The liberty-revolution, that was deemed a splendid success, became the blue-print to destabilize Egypt and later Syria, to depose their governments likewise. And since none of this helped to slow the financial collapse, the elected government of the Ukraine was violently overthrown by lavishly financed 'hired' Nazi 'revolutionaries' as a stage for war against Russia, and by alliance, also China. Any war against Russia, invariably becomes nuclear war that involves both, Russia and China, which adds up to a madness that is unsurvivable.

That's where we stand today, with the western financial system now almost totally dead on its knees, which not the greatest sacrifice in the world can revive.

To restore the Glass Steagall law

1933 - the Glass-Steagall law to protect that value of the U.S. currency - now repealed

A substantial political movement has begun in recent years to restore the Glass Steagall law in order to halt the train to hell. Obviously, if the law had not been repealed, the economic, financial, and strategic tragedy that the nation of the USA, and also the world, has suffered, would likely have been prevented. However, reinstating the Glass Steagall law at the present stage is no longer sufficient. Too much has been destroyed on many fronts that storing an old compromise would solve the crisis that has become a national tragedy.

Solving the tragedy at this stage

Solving the tragedy at this stage would require an uncompromised stand for the principle of anti-entropy being reflected in the universal self-development of humanity on the entire front of civilization. Nothing less would do. This would include anti-entropic finance and economics, large-scale scientific and technological development, national banking, national directed credit creation, quality education, health care, universal high-quality housing, and a commitment to uplifting culture. This complete shift to an uncompromised, anti-entropic standpoint on the entire front, would leave no room for stealing.

*To merely reinstate Glass Steagall, defies the nature of reason

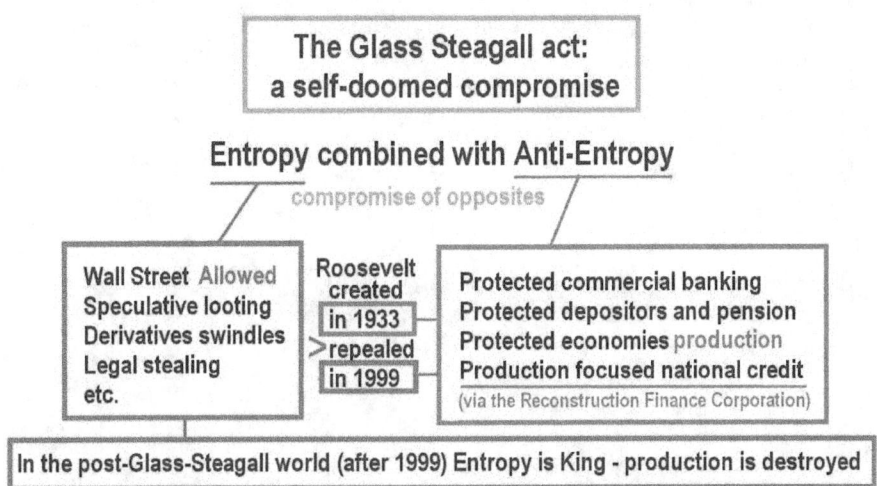

The Glass Steagall act:
a self-doomed compromise

Entropy combined with Anti-Entropy

compromise of opposites

| Wall Street Allowed Speculative looting Derivatives swindles Legal stealing etc. | Roosevelt created in 1933 >repealed in 1999 | Protected commercial banking Protected depositors and pension Protected economies production Production focused national credit (via the Reconstruction Finance Corporation) |

In the post-Glass-Steagall world (after 1999) Entropy is King - production is destroyed

The suggestion to merely reinstate Glass Steagall, defies the nature of reason, which is itself anti-entropic. For example, why would we restore elements of a failure that has collapsed the system itself, which the kingdom of entropy brought about, but which was tolerated under Glass Steagall? Or why would one bring back even the anti-entropic element of Glass Steagall that had been insufficient in itself to eclipse the entropic elements. It appears that the Glass Steagall compromise was made, because the principle of anti-entropy had not been developed extensively enough to be understood in 1933.

Ice Age 2050s: Certainty

This sets the stage for the surprising recognition that the platform for meeting today's vastly larger challenge, the Ice Age Challenge, must rest on a dramatically higher level than just getting back to an old bill that had failed.

To create a more-just economic order in America

1933 - the Glass-Steagall law to protect that value of the U.S. currency - now repealed

Glass Steagall had originally been designed to create a more-just
economic order in America, in which society is protected against
the ravishing rule of entropy. While this law has met the most
pressing needs of the time and has enabled unprecedented
prosperity and stability, it falls short of meeting today's
requirement for meeting the Ice Age Challenge where vastly greater
imperatives stand before us than the 'little' concerns that the Glass
Steagall Act had dealt with.

Seeing the True Sun

In order to meet the great challenge that is no longer avoidable, the self-rescue of society needs to begin with the principle of anti-entropy, which needs to overturn all previous laws, falsely written, that don't measure up to its standard. Humanity needs to do this all over the world, to be true to itself. Humanity, is the only anti-entropic species on Earth that has the capacity with its intellect to look at the universe, discover its principles, and with the discoveries leap ahead of illusions and notions and reach deep into the future and bring the future demands into the present for the shaping of policies in order that life can be preserved, and a rich civilization arise, three decades from the present in a radically altered world that no one in remembered time has ever seen or experienced, but which can be known in the mind.

Nothing is gained from clinging to the past

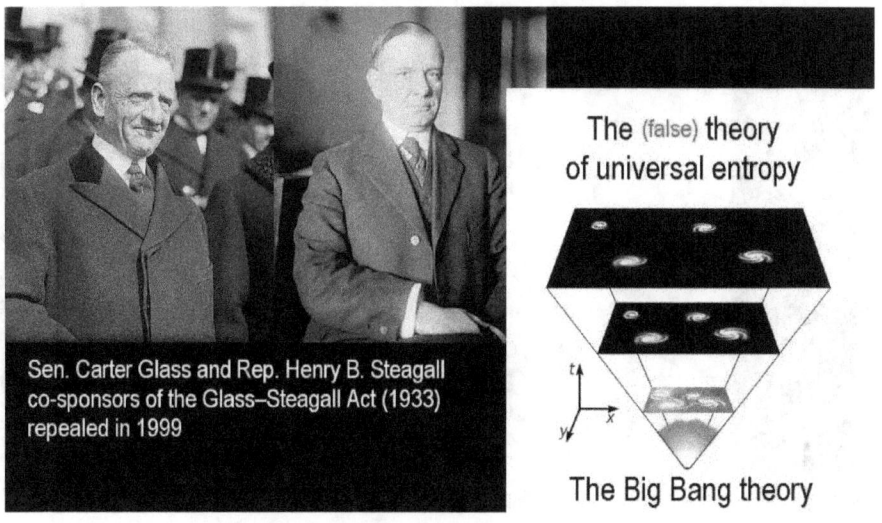

Sen. Carter Glass and Rep. Henry B. Steagall
co-sponsors of the Glass–Steagall Act (1933)
repealed in 1999

The (false) theory
of universal entropy

The Big Bang theory

Restoring an old platform from the past that has failed, whether it is political in nature like America's Glass Steagall banking law that had enabled America to become the most prosperous nation in the world, but which has failed, or whether the platform from the past is scientific in nature, such as the more obviously failed Big Bang theory, any such clinging to the past is ultimately an exercise of futility that in effect celebrates entropy, the very sense of entropy that the false Big Bang theory places before humanity. Nothing is gained from clinging to the past, as inviting as this may seem, because to do so denies our power to live in the future, where we can look forward to greater achievements than any that have been wrought in the past.

By becoming latched to the past

By becoming latched to the past, people live by the entropic assumption that humanity doesn't have the capacity anymore to leap ahead into the future and pull the present up behind it.

By staying latched to the past

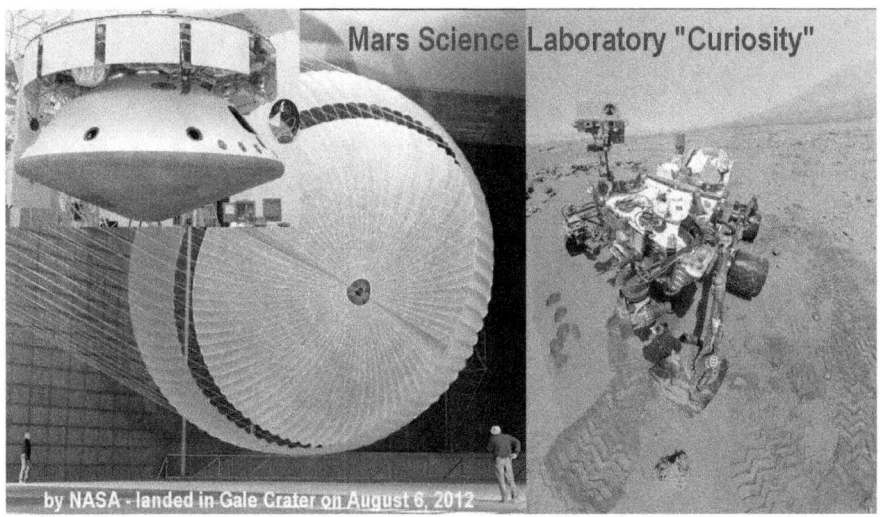

Mars Science Laboratory "Curiosity"

by NASA - landed in Gale Crater on August 6, 2012

By staying latched to the past, people lie to themselves. They lie to themselves about their own humanity, because it is knowable that each person as a human being has the capacity to know the truth. This applies in science, economics, politics and in civilization as a whole

Without the advancing recognition in society

Pudong International Airport
Shanghai
China

Mang Lev
431 km/h
top service speed

Without the advancing recognition in society of the anti-entropic
nature of the human being, which no other forms of life on earth
can match, humanity would doom itself, both in the present and in
the future, because our very existence depends on our human,
ever-expanding, creative capability, and increasing scientific and
technological progression.

Glass Steagall has become too shallow

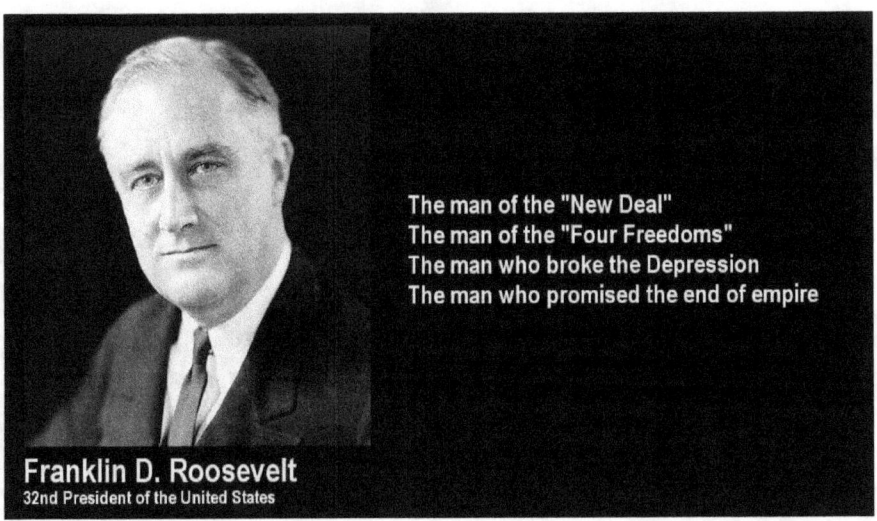

The man of the "New Deal"
The man of the "Four Freedoms"
The man who broke the Depression
The man who promised the end of empire

Franklin D. Roosevelt
32nd President of the United States

This means, that the imperative that the Glass Steagall act in 1933 had been based on, on the political front to advance America's economic progress that was much needed 65 years ago, has become too shallow to make the grade in today's world. We need to go further, by a long way. Compromises, no matter how tempting they may be, are no longer useful, nor beneficial. Any compromising on the grand scale of civilization has become too dangerous for the whole of humanity, because the imperative in today's world is that we move ahead on all fronts without fail, because the Ice Age Challenge cannot be met with anything less.

The Need for Looking Forward

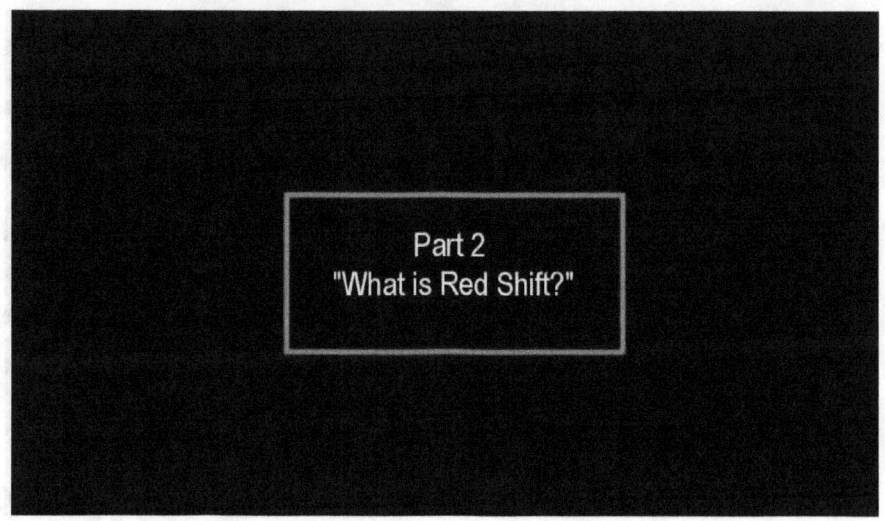

"The Need for Looking Forward"

The imperative comes with the Ice Age Challenge

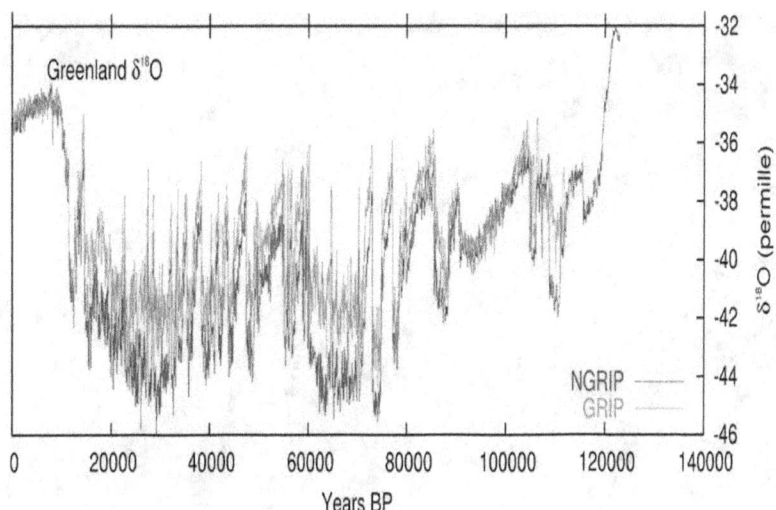

The imperative that inspires us to become more human, comes with the Ice Age Challenge that cannot be side-stepped or be compromised with. Nothing less will do, than moving ahead as effectively as possible.

It challenges us all to become human beings

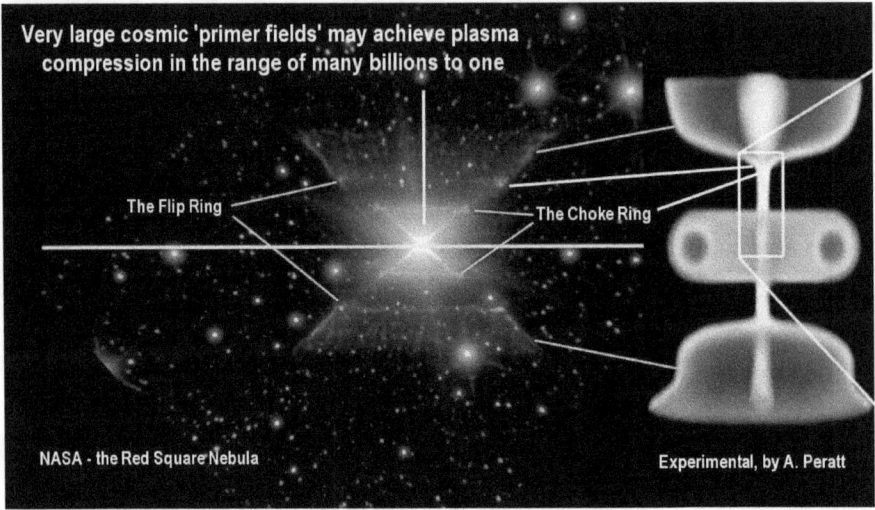

It challenges us all to become human beings in the fullest sense of our highest self-recognition, and that we move with the universe by utilizing its resources, and the resources we have within as children of the universe, who have the power to become one with the universe. The demand, that we place on us, takes us far beyond the falsely imagined Big Bang entropy.

To create an Ice Age Renaissance

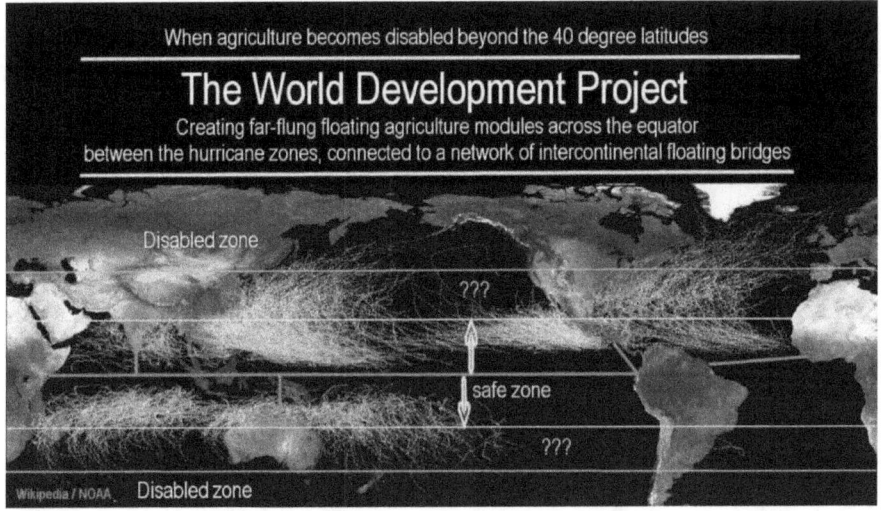

We place this demand on us, for the obvious reason that the Ice Age Challenge can not be met with less than the complete commitment in society to the principle of universal anti-entropy that is reflected in humanity worldwide, just as it is prominently apparent in the operation of the universe, and on earth in every form of renaissance that ever was. The dynamics of the universe challenge us, to create an Ice Age Renaissance for ourselves, on a scale that far supersedes our grandest dreams to date, because we have become accustomed for far too long, to dream too small.

We need to look forward with the eyes of science

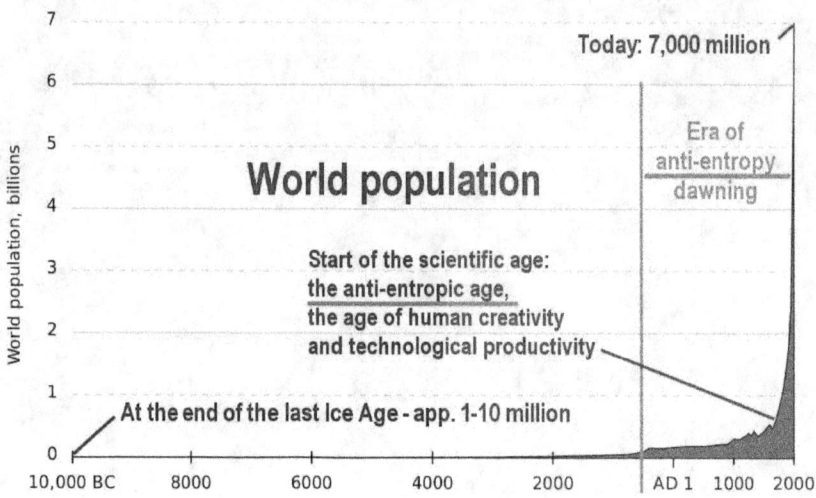

The point is that we need to look forward with the eyes of science, and consistently step beyond the platforms of the past and their limits and failures, even as we built on past achievements and experiences.

In terms of our natural capacity as human beings

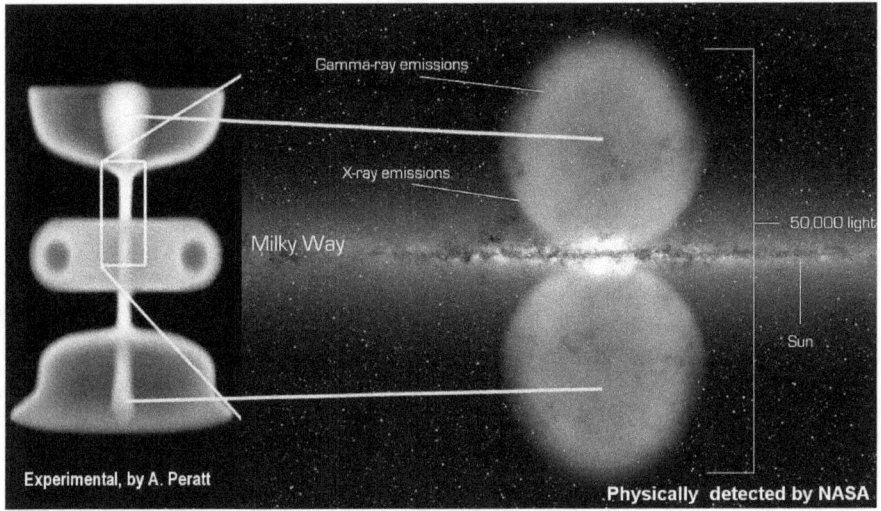

This means that we focus on what is real in the universe and in ourselves as human beings as a part of the universe, and that we use this higher-level focus as a starting point.

In terms of our natural capacity as human beings, we are presently operating at a large deficit on the road towards meeting the Ice Age Challenge. Too many centuries have been wasted under the chokehold of entropy in the kingdom of thievery, such as the Roman Empire and the like, have placed on us.

The Ice Age phenomenon is not the product of entropy

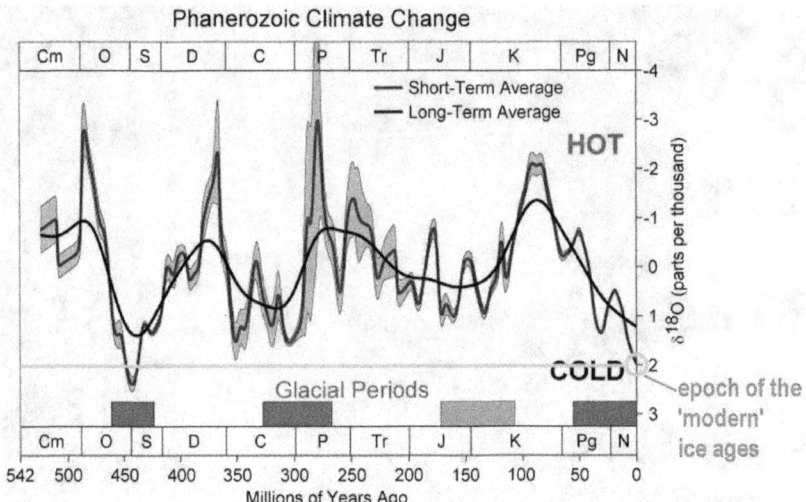

The Ice Age phenomenon by itself, is not the product of entropy. It is a phase of the anti-entropic cycles of the universe, and of ever-higher forms of life unfolding within them.

Children of the anti-entropy of the universe

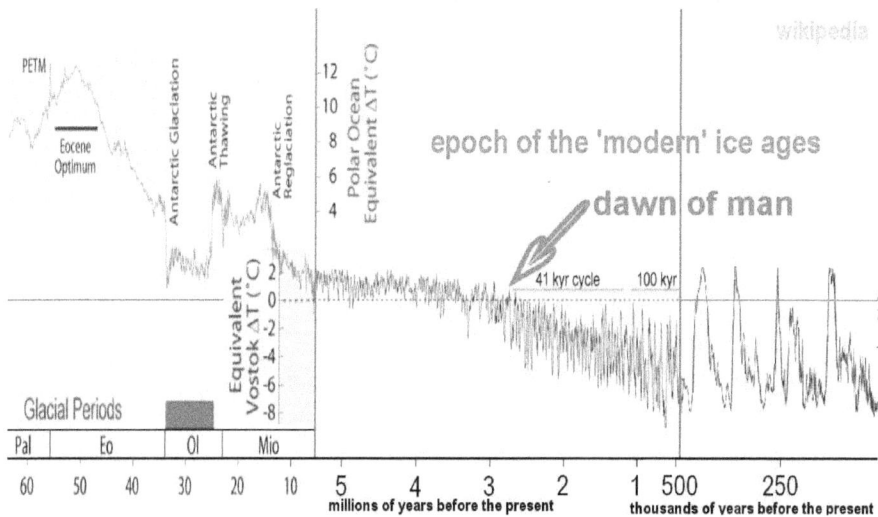

We, ourselves, as intelligent human beings, are the children of the anti-entropy of the universe that unfolds evermore forms of beauty, sublimity, and power, almost in its own image.

It is highly likely that the dawn of humanity on Earth had not been possible on earth until specific astrophysical conditions had existed that had enabled a breakthrough in the development of a higher-level species, and by several successive breakthroughs that have occurred subsequently, which would render us the children of the universe.

We may fear the Ice Age, but in real terms the ice ages have been our cradle in which substantially larger volume of galactic cosmic-ray flux have reached the Earth for most of its time, which didn't happen before our time.

We, as humanity, emerged from the cradle of the ice ages as the brightest stars in the heavens of life, and with a potential direct connection with the universe itself, a connection that we have just begun to discover.

According to all evidence, we are on a track that takes us far beyond the notion of entropy. We are not glorified animals, we never have been, but are the diamonds in the heavens of life.

Ice ages are critical elements in the progressive dynamics

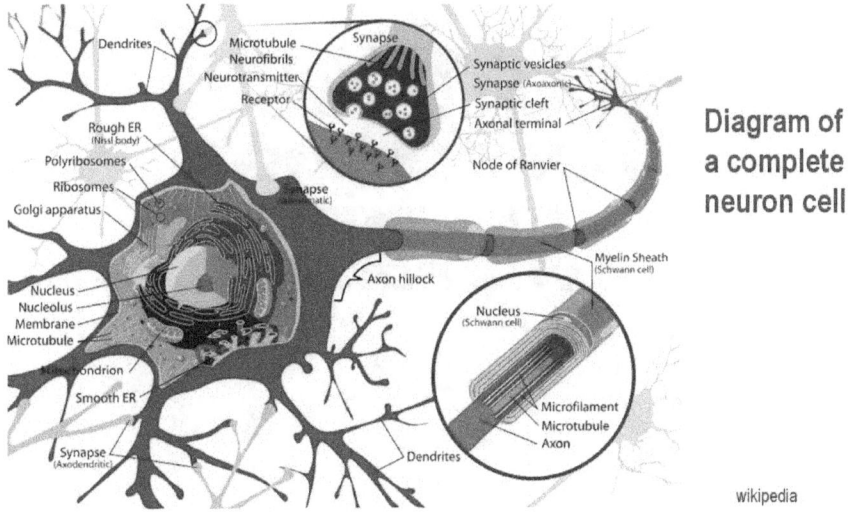

Diagram of a complete neuron cell

The cosmic-ray flux that has affected us richly through our past, is an electric phenomenon, a phenomenon of fast moving electric particles, and related particles. This fast electric flux has the potential to be a critical factor for the development of complex neurological systems as we have within us that have become systems of cognition, and ultimately, consciousness. In this context we may be the children of the creative, anti-entropic, quality of the universe in the most intimate sense, where the ice ages are critical elements in the progressive dynamics.

The Big Bang theory stands plainly in denial

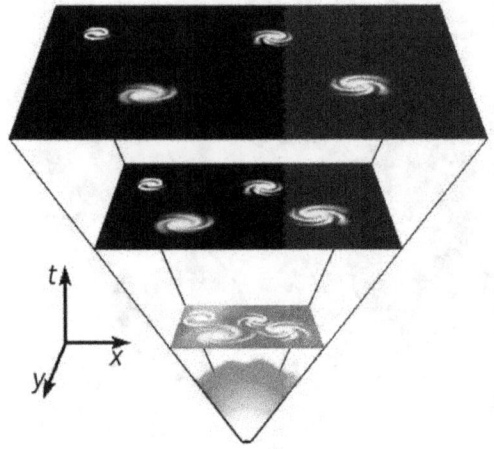

The Big Bang theory stands far, far distant from the anti-entropic principle of the universe, though it took us a long time to discover the principle that is expressed in our humanity.
The discovery that we have made on so many fronts, tells us that the Big Bang theory stands plainly in denial of what is self-evidently true.

Big Bang theory might have been intentionally staged

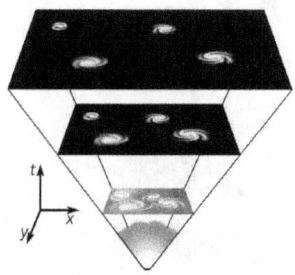 

While no impelling evidence exists that the astrophysical Big Bang theory had been intentionally developed for political purposes, and might have been promoted as a cultural warfare project to create an empty center in society - in science, economics, and in culture - the timing of the promotion of the theory - as a counter-ideology - seems to suggest that the Big Bang theory might have been intentionally staged for such intentions.

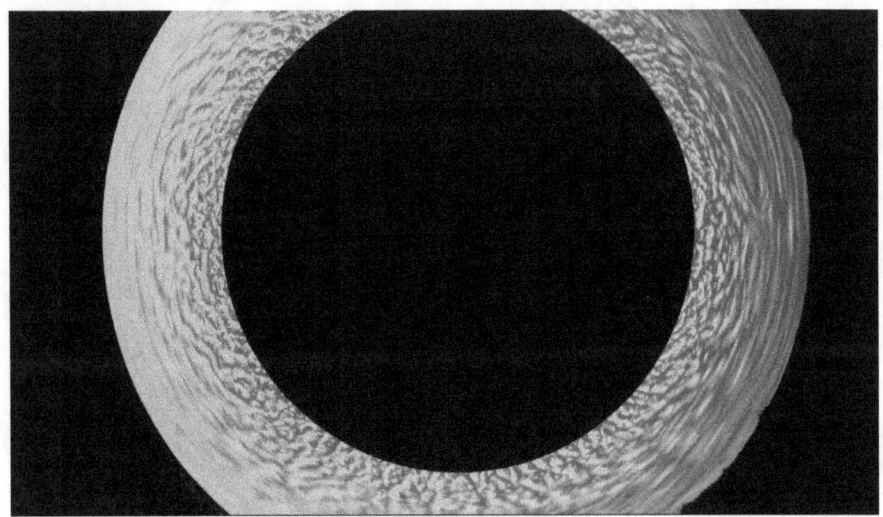

This factor, all by itself, should inspire us to regard the Big Bang Cosmology and its fire of entropy with an empty center, as nothing more than a tragic product of false assumptions at best, and false intentions at worst.

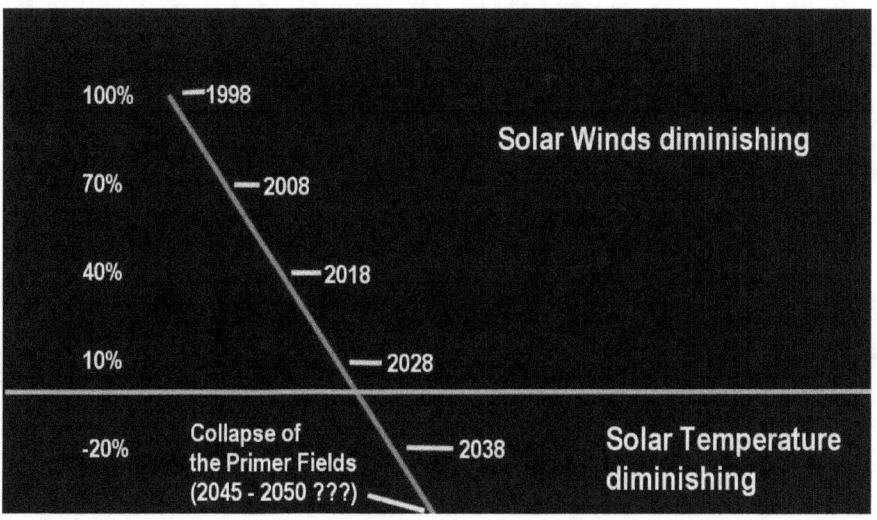

The Ice Age is near, as near as the 2050s, potentially. With the challenge that this poses, in mind, it shouldn't be too hard for society to step up to higher ground and move forward from there to build itself a correspondingly great renaissance, on the basis of this truth, with which to rebuild its deeply-collapsed civilization in order that the Ice Age Challenge be met in time, before the physical Ice Age starts anew.

A great need for a renaissance of truth

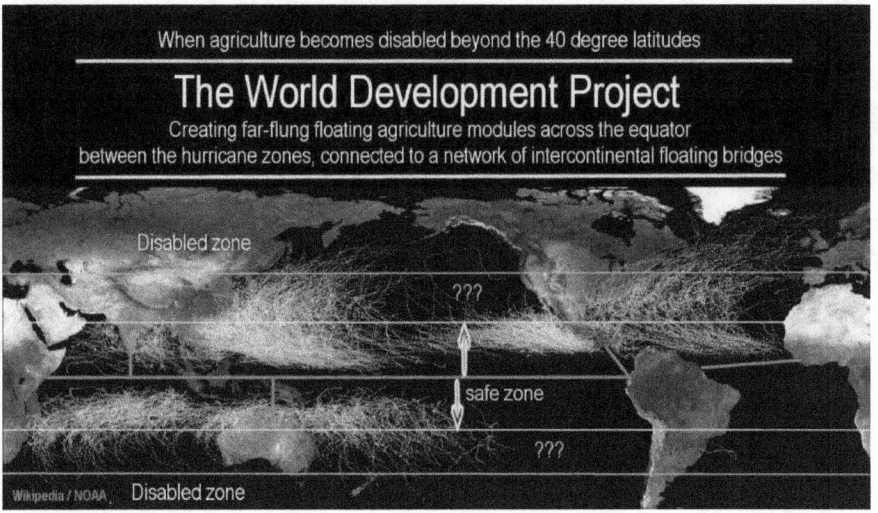

We have a great need for a renaissance of truth towards this end, because the truth is, after all, the builder of worlds, and truth does not diminish.

The truth is here to stay, and the truth is, that humanity is an anti-entropic power in the world. For this, we have ample evidence. In truth we find the foundation for our future. "In Truth We Trust." That's what the new banners proclaim.

The proof of the 'pudding' is unmistakable

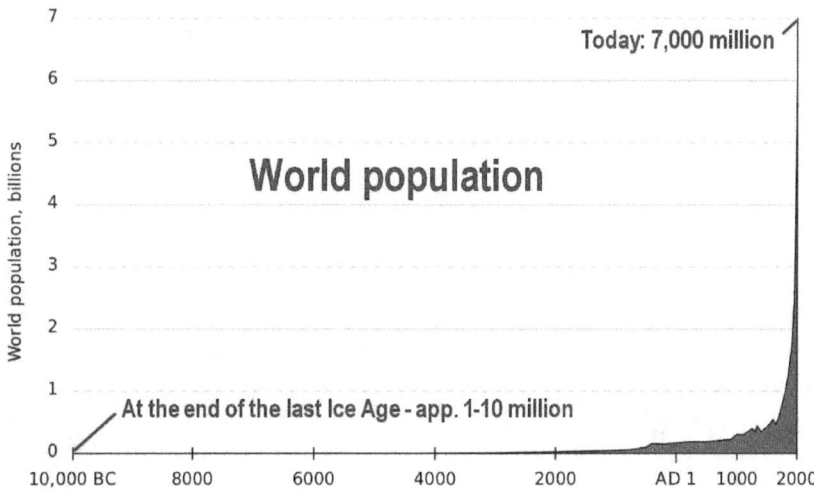

In spite of the millstones that have been hung around humanity's neck during the hellish period of its history, of the kingdoms of empire, humanity has forged ahead to ever-greater levels of self-development, in science, technology, and humanity. The proof for this is amazingly evident in the 'pudding' of increasing population density that has been achieved.

The proof of the 'pudding' is unmistakable. Until scientific thinking began to dawn in the world, at around 300 BC, the world population had remained almost stagnant. Then with the scientific methods of Socrates and Plato setting a higher stage for humanity, a new wind began to blow that opened the horizon of the mind.

the sky is no limit" or "there are no limits."

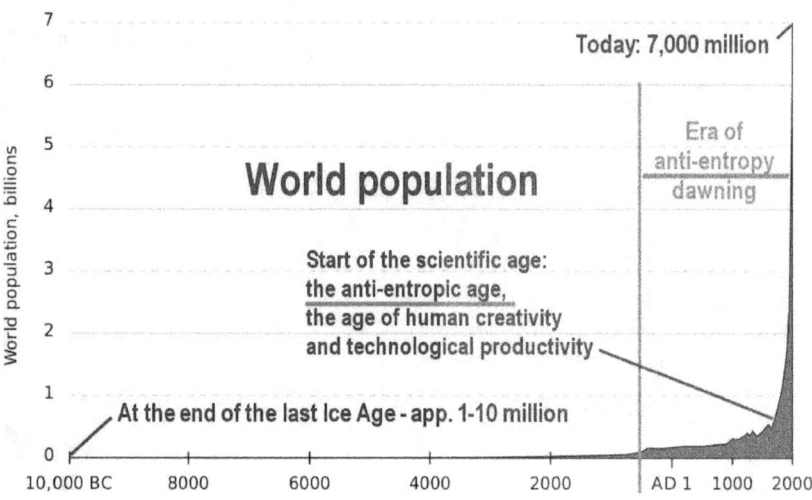

Suddenly, with small-minded thinking becoming put aside, a dramatic increase in economic power began to develop. The increase was slow at first, choked by the devastating effects of the continuing empires of the world. The chokehold was broken, however, by the new breakthrough in science that unfolded with the work of Johannes Kepler, who started a new freedom in thinking and in humanity's self-perception.

As the result, society became more productive, more innovative, more creative, as if the watchword was spoken, "the sky is no limit" or "there are no limits."

With increased industrialization

With increased industrialization, and advances in farming, mechanization, transportation, and so on, in an environment of increased energy use, it became possible for evermore people to support themselves on the lands of the Earth that in primitive times supported just a few.

The sharp increase in the human world population to the 7 billion level in our time, was obviously not the result of improved breeding habits, but does simply reflect the anti-entropic power of humanity becoming increasingly recognized, with which society creates itself new resources for living with advanced technologies, and so on, all provided by the anti-entropic human intellect.

Population increase mirrors anti-entropic economics

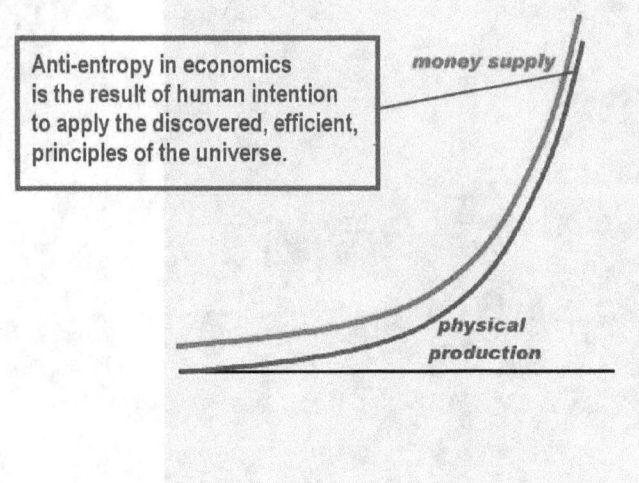

The worldwide achieved population increase mirrors the natural dynamics of anti-entropic economics.

The graph shown here illustrates merely the principle of the dynamics. The principle is addressed in Segment 2 of the video series on the Big Bang entropy theory.

The rate of increase that can be achieved

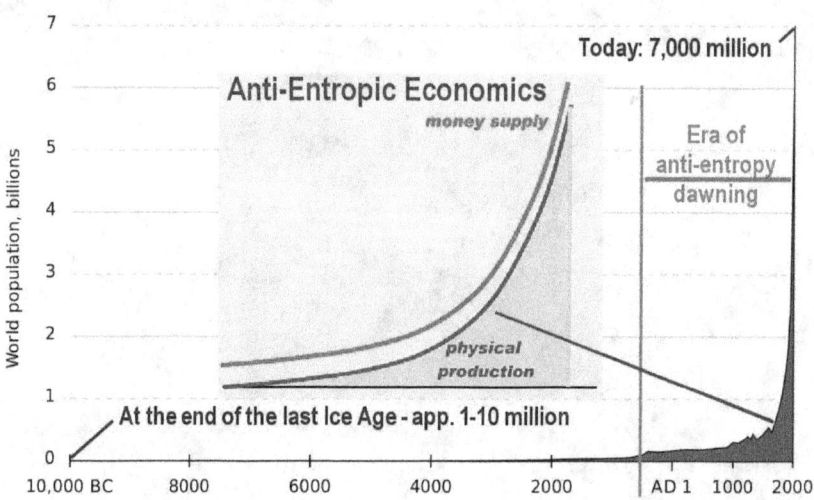

The rate of increase that can be achieved, of the expression of the dynamics, is only limited by the limits society is placing on the human spirit in its surging ahead.

Floating bridges across the tropical seas

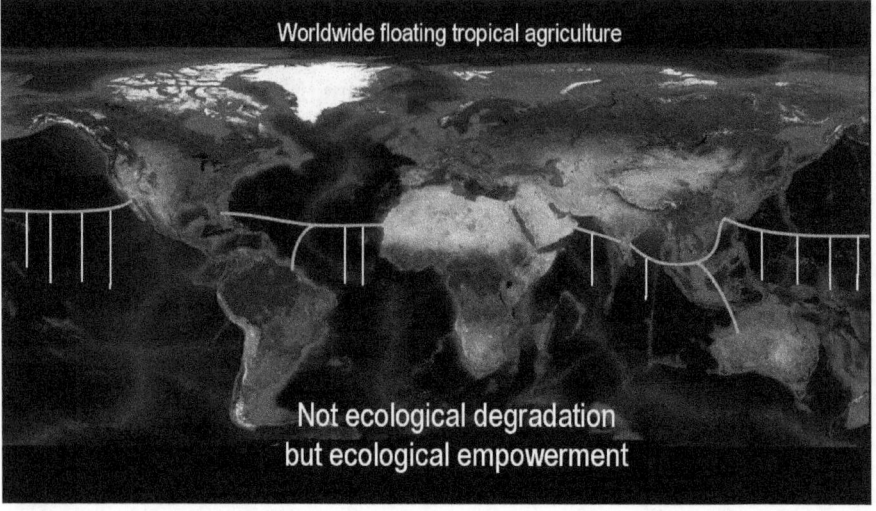

Worldwide floating tropical agriculture

Not ecological degradation
but ecological empowerment

What would prevent humanity today from laying down floating bridges across the tropical seas that connect up with floating agriculture that will take over the food production for the world when the recurring Ice Age disables agriculture outside the tropics, potentially in the 2050s timeframe.
We have the materials for it in the form of basalt, and the energy resources for it in the form of thorium nuclear fission. We have the technology already in use on a small scale. Once we can get our spirit roused to do this, this gigantic seeming project will become a small thing

The 6000 new cities for a million people each

We would likely place most of the 6000 new cities for a million people each, which we will need, afloat onto the seas, together with their agriculture, in preparation for relocating the nations out of the northern areas that become uninhabitable in an Ice Age environment, and we would create the cities for one-another for free. With automated, high-temperature industrial production, the mass-production of quality housing will become so easy that cost-free living will become the new basic infrastructure for humanity meeting its human needs in the most-efficient manner possible, as an open door for further advances in scientific and cultural achievements. All of this becomes possible when we lay aside the entropic, small-minded, mode of self-perception, to an open-ended perception.

To break down the barrier in the mind

Water
the blood of life

Our greatest need therefore, is to break down the barrier in the mind that would prevent us from meeting our potential as human beings. The barrier of small-mindedness is already killing far too many people with poverty, even in the richest of the rich nations. With the increasing drought conditions, as a fringe effect towards the coming Ice Age, the USA is fast running out of freshwater resources. The shortage could have been prevented.

Drought conditions can be offset

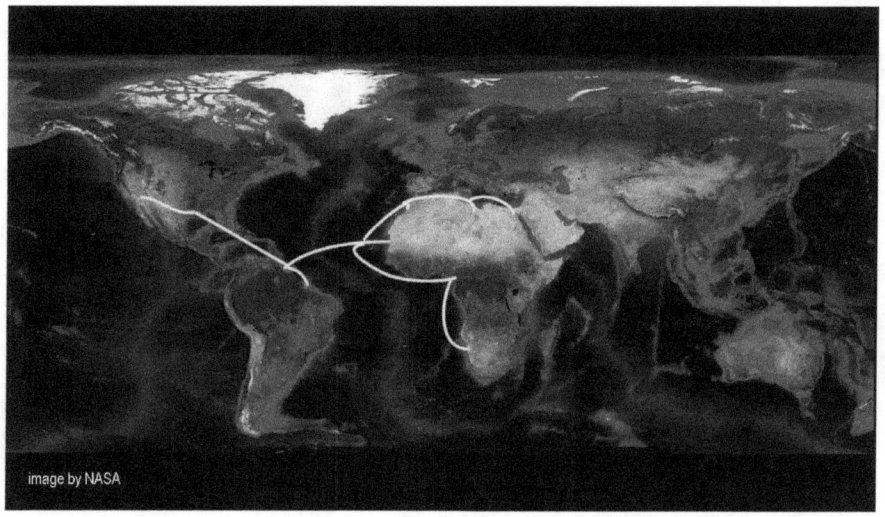

image by NASA

Drought conditions can be offset by simply redirecting some of the outflow of the great tropical rivers, such as the Amazon River, to the dry areas of the world via a network of floating arteries made of woven basalt.

Desalination of ocean water provides an endless resource

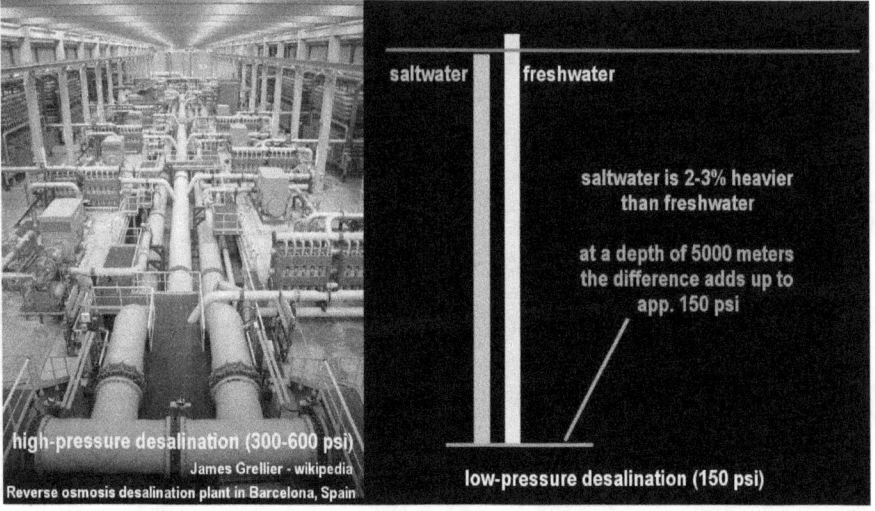

saltwater   freshwater

saltwater is 2-3% heavier than freshwater

at a depth of 5000 meters the difference adds up to app. 150 psi

high-pressure desalination (300-600 psi)

James Grellier - wikipedia
Reverse osmosis desalination plant in Barcelona, Spain

low-pressure desalination (150 psi)

Actually, we don't need to rely on rivers at all, in order to meet our freshwater needs. Desalination of ocean water provides an endless resource. All we need to do is utilize the weight differential between freshwater and salt water, to self-power deep ocean reverse osmosis desalination systems, to cause rivers of freshwater to flow out of the oceans, independent of weather conditions. The technology already exists. The deep-ocean application hasn't been implemented yet, because small-minded, entropic thinking that accepts the collapse of human living as normal, has prevented the application of the humanist power we already have at hand.

Recognized already during the Kennedy era

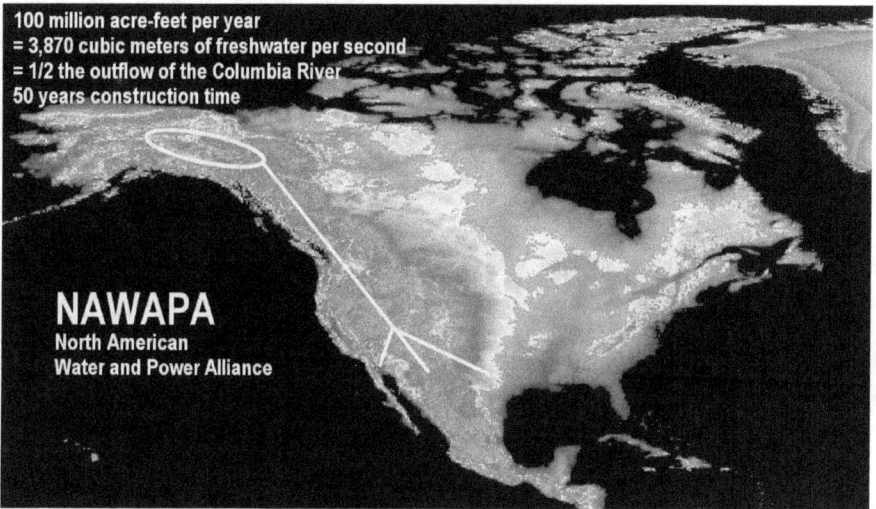

100 million acre-feet per year
= 3,870 cubic meters of freshwater per second
= 1/2 the outflow of the Columbia River
50 years construction time

NAWAPA
North American
Water and Power Alliance

It was recognized already during the Kennedy era that the south-western dry areas of the USA could be made far more productive with increased volumes of freshwater. For this reason a project was devised to divert portions of the large northern rivers to the south for increased agricultural production. The giant NAWAPA project was never even started. Now, with the drought setting in, the Southwest is in a water crisis that is decimating agricultural production in the region, instead of increasing it.

Deep ocean reverse osmosis desalination

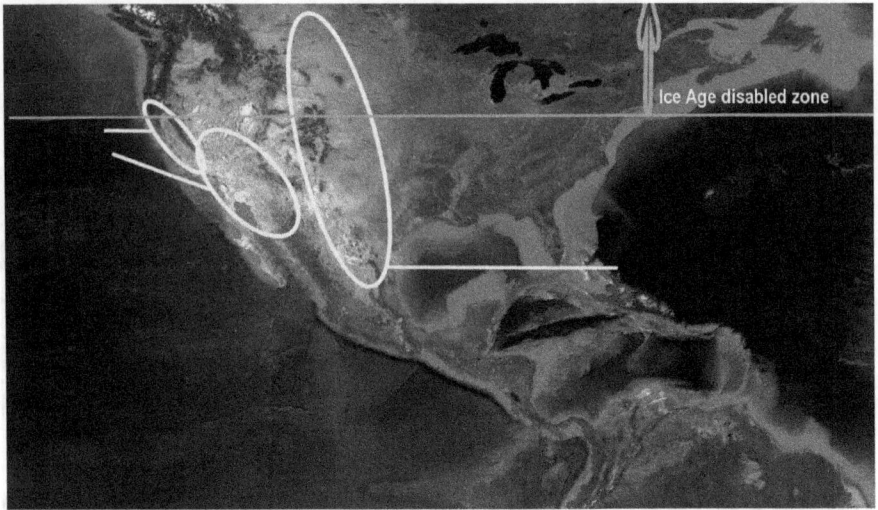

While it would take a 50-year construction effort to divert water from Alaska, deep ocean reverse osmosis desalination can be built relatively easily. The deep oceans exist. The volume of freshwater that can be produced depends only on the size of the infrastructures that we built to meet our needs. And those needs will increase, dramatically, especially during the coming Ice Age environment when rain becomes scarce. Desalination becomes essential then, and not only for agriculture.

Deep ocean reverse osmosis desalination

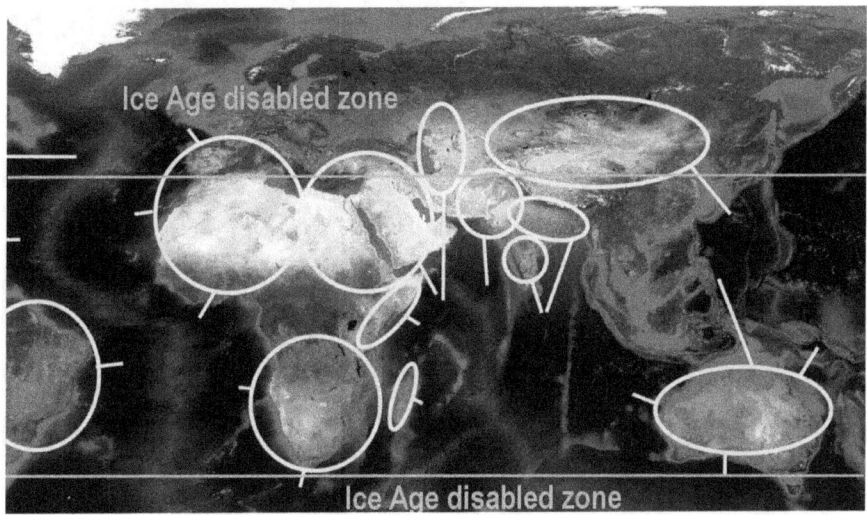

The entire world will find itself in the same situation. It has the same opportunity for unlimited freshwater production from deep ocean reverse osmosis desalination that America has available to it. Deep oceans are nearby everywhere.

India can never suffer water shortages

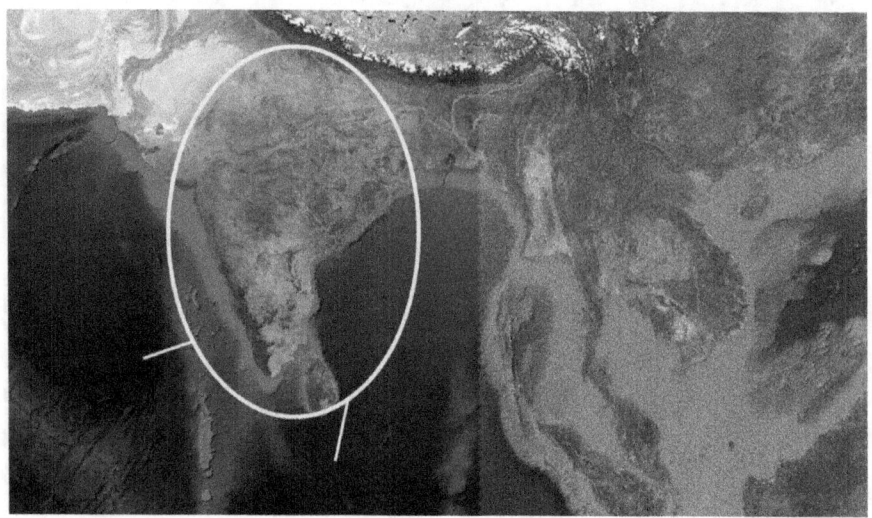

India can never suffer water shortages if the infrastructures are being built. Those will become critically essential for the coming Ice Age when the rain becomes scarce. This means that the infrastructures should be started now, including the inland distribution systems. Large infrastructures are not built quickly, especially when the industries for building them need to be created first. It is tempting to say, that these gigantic efforts are not possible, but in saying this we would be denying ourselves. We would deny us as an anti-entropic species. With this denial we would be laying ourselves down to die.

*The relocation of entire nations becomes essential

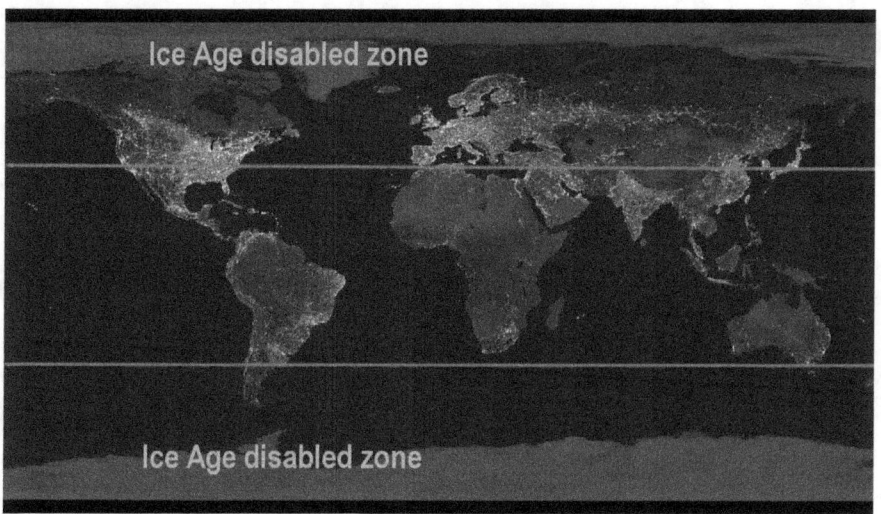

The tropical world, in which the future of humanity will have to be located in 30 years when the Ice Age begins anew, is presently vastly underdeveloped, while the highly developed regions are located in the zones that become uninhabitable. The relocation of entire nations becomes thereby essential. This means that enormously large projects will become common, because it is unthinkable that humanity will simply lay itself down and commit suicide by default. This won't happen.

We will built the 6000 new cities

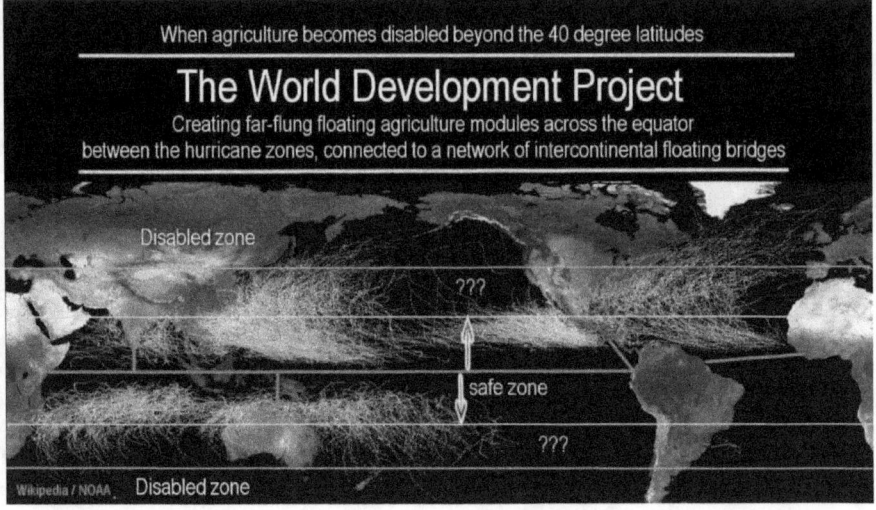

When agriculture becomes disabled beyond the 40 degree latitudes

The World Development Project

Creating far-flung floating agriculture modules across the equator
between the hurricane zones, connected to a network of intercontinental floating bridges

Sure, some decisive efforts will need to be made to get us all out of the present slump of small-minded thinking that keeps us tied to a decaying landscape where impotence and entropy rules, especially economic entropy that invites stealing, where we say this is too hard to do, who will pay for it all, so that nothing is presently being built. However, the prospect isn't pleasant either, that when the Ice Age begins in the 2050s and devastates agriculture outside the tropics, that most of humanity will starve to death, because people don't live long without food. Consequently, we will stir our stumps and get the job done, to meet our future needs. This means that we will built the 6000 new cities for a million people each and the millions of acres of floating agriculture that are required, because this is the human thing to do, and an exceedingly exciting thing to do as well. Most likely, we haven't seen anything yet in terms of what we can accomplish as human beings with our profound intellect and creative capacity. This will set the stage for the future.

Whether we survive the Ice Age Challenge with vast new infrastructures in place in the tropics in preparation for a potential Ice Age Renaissance world, or whether we will fail ourselves and die of starvation as a consequence, depends exclusively on the recognition of ourselves as an anti-entropic species that inspires us to accomplish the needed tasks.

We have ample of proof to our credit

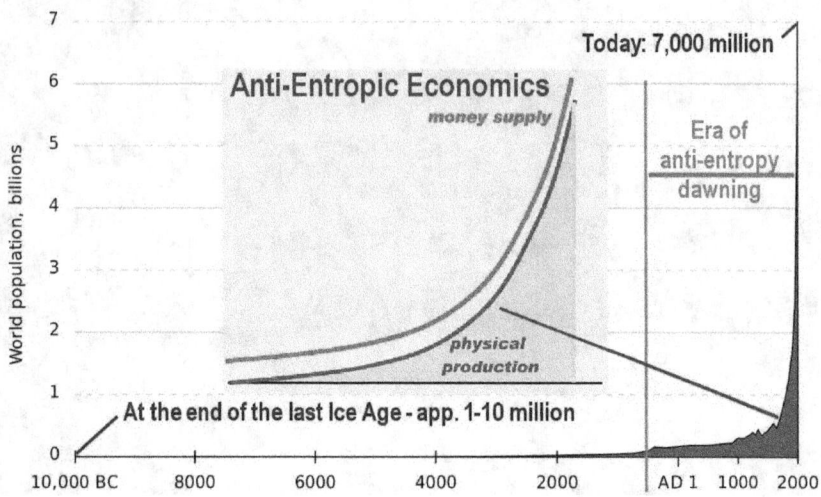

I would like to suggest that we will not fail ourselves, because we have already demonstrated to ourselves that we are vastly more capable than we presently give ourselves credit for, for which we have ample of proof to our credit.

Entropic factions in politics and philosophy

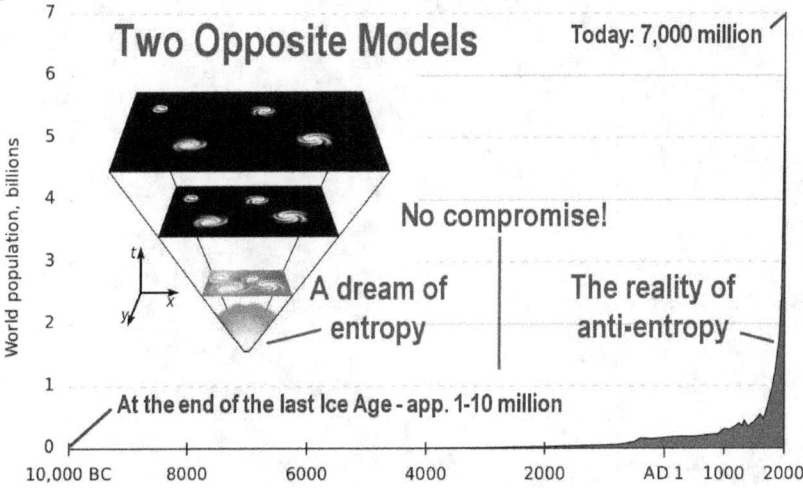

While entropic factions in politics and philosophy see the amazing proof of the human anti-entropy with fear, and say that humanity is overgrazing the land that will collapse the biosphere and thereby diminish the resources, they speak from ignorance and self-dilution. Their fear is false, built of false historic philosophies that are as false as the Big Bang theory is false that appears to be intentionally entropic.

The reality is that nothing is winding down anywhere in the universe, nor is the potential of humanity winding down. Our open-ended development potential has been demonstrated as truth, and the truth does not diminish. In fact we have just begun to peck open the shell of our infancy, reaching for infinity.

As the truth is being experienced, we begin to fly high

Photo by Scott Williams

The Supreme Being

wikipedia

In discovering our truth, by the discovery of the creative principles
of the universe, we discover ourselves as the supreme being on
Earth in the living image of God or the Universe.

Then, as the truth is being experienced, we begin to fly high above
the dust of entropy and find that our world has become cleansed of
all of its pesky, numerous, small-minded illusions, which thereby
has become transformed into fruitful fields and gardens of beauty
and happiness. And for that, we have the means already at hand.

www.ingramcontent.com/pod-product-compliance
Lightning Source LLC
Chambersburg PA
CBHW060402190526
45169CB00002B/713